HSC Maths
Introductory Calculus

Basic Differentiation & Integration

Graham & Kathleen Fandouly

A
FIVE SENSES
PUBLICATION

Five Senses Education Pty Ltd
2/195 Prospect Highway
Seven Hills 2147
New South Wales
Australia

First Published 2004

Fardouly, Graham and Kathleen,
Top Notes –Introductory Calculus
ISBN 978-1-74130-172-4

TOP NOTES (2 UNIT) MATHEMATICS SERIES

This series has been created to assist Preliminary and HSC Mathematics students in their understanding of these courses. It begins with a thorough review of the basic skills required to achieve success in Mathematics, then steadily and thoroughly covers all the topics contained in the syllabus.

The first two books mentioned in the following list introduce students to the **arithmetic**, **algebra**, **functions** and **graphs** that underpin the central concept of this course - **calculus**.

The basic concepts and skills of differentiating and integrating functions are included in the important **Introductory Calculus** book, while the three main **Calculus** books specifically cover all the applications.

Trigonometry focuses on the range of non–calculus skills included in this topic. The remaining books in the series cover topics that are not necessarily directly related to calculus, but are still very important in their own right – **Geometry**, **Probability** and **Series**.

In each of these books, the reader is allowed to develop a sound understanding of the main ideas through clearly explained examples, then offered an opportunity to apply these skills through appropriately graded exercises. Fully worked solutions are provided for all questions, enabling students to immediately mark and correct their work.

The books in this series are:

- Arithmetic and Algebra
- Functions and their Graphs
- Introductory Calculus – Basics of Differentiation and Integration
- Calculus I - Geometrical Applications of Differentiation and Integration
- Trigonometry
- Geometry – Plane and Coordinate Methods
- Calculus II – Trigonometric, Exponential and Logarithmic Functions
- Calculus III - Applications to the Physical World
- Probability
- Arithmetic and Geometric Series

CONTENTS

WHAT'S IN THIS BOOK?

This book contains:

- A general course description.
- Some very useful ideas about how to study for success in Mathematics, and the language skills required to fully grasp the basic concepts of the course.
- A list of all subtopics associated with the syllabus topics, *The Tangent to a Curve and the Derivative of a Function, and Integration.*
- An easy to read summary and worked examples for each subtopic.
- A sensible number of exercises for practice, with full solutions.

We believe that you will find these notes very useful in your study of the Mathematics course

Graham and Kathleen Fardouly

COURSE DESCRIPTION

The Mathematics course is divided into two parts – Preliminary and HSC. You will find questions from both sections in the HSC exam.

PRELIMINARY COURSE:

- Basic Arithmetic and Algebra
- Real Functions
- Trigonometric Ratios
- Linear Functions
- The Quadratic Polynomial and the Parabola
- Plane Geometry
- Tangent to a Curve and Derivative of a Function

HSC COURSE:

- Coordinate Methods in Geometry
- Applications of Geometric Properties
- Geometrical Applications of Differentiation
- Integration
- Trigonometric Functions
- Logarithmic and Exponential Functions
- Applications of Calculus to the Physical World
- Probability
- Series and Series Applications

TEN STEPS FOR SUCCESS IN MATHEMATICS

1. Do Something

There are many and varied ways to study for success in Mathematics. The one factor common to all these techniques is just to **do something**.

Some students just read over their notes and call this their study and, for a very lucky few, this approach is fine. Unfortunately, the rest of us need to do more in order to achieve any measure of success in this subject.

There are a variety of effective ways to study. Try as many as you can and see which one, or ones, work best for you. But remember, if you are not spending enough time doing something related to the subject, you are giving yourself little chance of improving your skills and understanding.

In the pages that follow, we have set out nine more steps that have led many of our students to success in this course over the years. You might find the techniques you currently use here. If you do, that's great, but also remember that it is important to vary your study activities to gain the greatest benefit from them. Take note of all the suggestions and incorporate as many of them as you can in your study. You might be pleasantly surprised by the outcome.

2. Attend Class

Make sure that you **attend classes** as regularly as you can. If you must miss any, make sure that you quickly and accurately copy any notes you missed.

The cheapest and best way to learn mathematics is to listen closely to people who have a good understanding of the concepts you want to learn. For most of you, your teacher is the main, if not the only, person you know who actually has some idea of what's going on. They can be contacted relatively easily and, from our experience, are more than willing to offer extra assistance to motivated students.

If you have tried this option but require further help, you may have to pursue other avenues. However, your teacher should always be your first point of contact.

If your attendance is inconsistent, or you are not up to date with the classwork or homework, you are wasting a simple and inexpensive opportunity to acquire the knowledge and skills needed to succeed in this course. It requires limited effort to copy notes from a friend's book, but the results of such a simple activity can sometimes be quite amazing. You will find that as you get more and more up to date, the lessons you attend become more comprehensible and, as a result, more interesting and rewarding.

3. Rewrite Notes In Your Own Words

Now that you are up to date with the current work, it's time to perfect the required skills and develop a deeper understanding of the ideas involved. The best way to do this is to **rewrite** any theory notes **in your own words**.

To do this properly, you must first have a pretty clear idea of the concepts involved. If you're not sure about something, ask your friends for help, or go back to your teacher and ask them.

Writing, and rewriting, your own notes gives you the opportunity to think relatively deeply about the concepts you meet in this course. It enables you focus on the language of mathematics and, the better you understand the language that mathematics is written in, the more likely you will be able to comprehend the questions you encounter, and respond appropriately to them.

Finally, it is much easier to study for exams using notes **prepared by you, for you**. They are notes that are written in a language you can understand. When you return to these notes in later study and in preparation for exams, they seem that much more easy to read and follow. You will be pleasantly surprised by the many benefits that follow from this simple, yet effective study technique.

4. Prepare A Summary

When you finish a topic, don't just forget about it. Just as it is a good idea to rewrite your lesson notes for each day, it is also strongly advisable to **prepare a summary for each topic**.

Summaries can be written in a number of different ways. Some students like to use point form while others cover their page with boxes of information or diagrams. Look for a technique that works best for you. Most importantly though, remember to write your summary **in your own words**.

Finally, once your summary is finished it should not just be filed away in a folder, never to be seen again, but reviewed regularly. It must be read, rewritten and updated as often as possible, as more and more examples and applications are encountered.

In this way, your summary becomes a simple, yet accurate and thorough, record of everything you know about a particular topic up to that time. Summaries are a powerful tool that enable you to demystify many of the concepts met in the course.

Many of our weaker students have incorporated this technique into their study, and made steady improvement as a result.

5. Practice Makes Perfect

Probably the most important study suggestion we can make is for students to practise regularly. Every one of you will gain some benefit by practising your skills. However, the greatest gains are made by those who practise with a purpose.

The practice done in class, where the teacher explains a concept or skill and asks students to complete similar questions on their own, is one type. Some students believe this is all the practice they need to undertake in order to fully grasp the work involved. Obviously, this is not the case.

Practising questions related to a topic you have just finished summarising is a great way to test your current level of skills and understanding. To attempt questions when not fully prepared can lead to a decrease in confidence and motivation, two very important prerequisites for success in this subject.

Practice questions may range from the simple recall type to the more challenging, problem solving exercises. The larger the range the better. Appropriate questions can be found in numerous places – textbooks, study guides and past exam papers, to name but a few.

This series includes a lot of standard, original and challenging questions and we congratulate you on locating it, but don't just stop here. By including a large number of questions in your study, you are developing the ability to successfully respond to a range of questions, and constantly reviewing and improving your skills in the process.

6. Past Test Papers

Past test papers are another very useful resource. However, they need to be used appropriately to be effective.

The practice of sitting down for three hours and working through an entire past HSC exam under test conditions can be a good indicator of your current skill levels but should only be done sparingly and at the appropriate time. If you are not fully prepared for this challenging task, you are likely to fail to achieve your desired level of success and your confidence may be negatively effected.

You cannot fully prepare yourself for an HSC exam until you have finished studying the entire course. A more appropriate way to use past exams in your study is to focus on questions related to only one or two topics at a time. In this way, you are gaining useful feedback in areas you are currently studying. Problem areas can be located and rectified in your study.

These smaller practice tests can be done under test conditions, but only for a set time period. Success in these tasks can increase confidence levels, so ensure beforehand that the questions selected are appropriate to your current level of skills.

As your confidence and skill level increases, the range and difficulty of the questions included in these tasks can increase. An emphasis on steady improvement is essential.

7. Fully Worked Solutions

Whatever resources you use to acquire questions for your study, make sure they come with **fully worked solutions**. This point is extremely important.

By comparing your answer with those given, you have an immediate marking system available. Specific problem areas can be isolated, enabling you to focus your energies on a skill or concept in your study that requires particular attention.

If the solutions come with a suggested marking scale, you also have the opportunity to grade your answer. Write your mark next to your response. Later, you can return to the same question and actually compare your marks!

Now, try to resolve the problem on your own. You might surprise yourself. Sometimes, a simple mistake has been made and you can go onto the next problem immediately. However, if the problem is a significant one and you cannot figure out how to resolve it, it should be written down and returned to at another time. **Don't leave problems unresolved. Ask someone for help!**

If you do find a good question but it only comes with a single answer, or no answer at all, don't just leave your answer unchecked, ask your teacher to mark it for you. I'm sure they won't mind.

8. Ask For Help

This is probably the most important step of all. **Ask for help when you need it**. Rather than spending hours pondering a single question or concept, ask an appropriate person for help. The problem may be cleared up relatively quickly, freeing up time for you to overcome further obstacles. We have already mentioned asking your teacher for assistance, but don't forget about others you can access - for example, older siblings or friends, other teachers, and even the HSC Advice Line, which operates for a period of time close to the HSC examination.

However, if the concern remains, don't lose heart. Look at it in terms of a positive learning experience. It is extremely unlikely that the difference between success and failure in this subject will hang on the resolution of one problem. So, seek out new challenges in your study and return to this particular one later. Often, the solution to a problem comes about when you are thinking about something else.

When the problem is finally rectified, it is very useful to take another look at your summary of the topic involved, and modify it accordingly. Include the question in your summary if you wish. You will be pleasantly surprised when you look back at these questions at a later date and realise that they are not very hard to do after all. You just needed the time and effort to develop the necessary skills and understanding required to solve them.

9. Challenge Yourself

When you feel confident that you have developed the skills to correctly answer the basic questions in a particular topic, it is time to **challenge yourself** by attempting progressively more difficult problems in your study.

Your responses to these harder questions can be good indicators of your current level of understanding. Write these problems in your summary, along with your answers and the full given solution, making clear notes on where you had any problems. These problem areas must be reviewed as soon as possible in your study.

Finally, try the question again at a later date and compare your response with your previous effort. If you have been employing the study suggestions mentioned above, there is every likelihood that this type of question is no longer the big challenge it was for you originally.

Find a new challenge and repeat this process. Your confidence, enjoyment and marks should improve as a result.

Of course, if you are finding these problems too difficult, you will need to return to your summary. Review the basic concepts and skills again, and only go on to these harder questions when you feel ready to do so. Sometimes, a basic understanding of a topic has to be good enough.

10. Enjoy Yourself

Sometimes, a lot of study is required before any measure of success can be achieved in this course. So, try to **have some fun** in the process.

Create some activities that you enjoy doing. Colour code your notes or draw pictures or cartoons to help you gain a better understanding. Use your natural interests and skills to motivate yourself.

For most students, an enjoyment of mathematics comes naturally with success. Mathematics is tedious and boring if you don't understand it. The best way to break out of this negative cycle is to practise regularly and review consistently. You might be pleasantly surprised with the improvement that takes place.

With a positive attitude, and a willingness to ascertain which approach works best for you, each and every one of you is capable of learning to enjoy the challenges associated with Mathematics.

Success breeds enjoyment, and vice versa.

THE LANGUAGE OF MATHEMATICS

To achieve any level of success in Mathematics, you must first become proficient in the language in which it is written. The language of mathematics can be a strange one. Many words used in the subject are also used in the wider English language, but have different meanings.

A vast array of symbols and terms are used, and you will need to know what each of them actually mean before you can fully understand what a question is asking you to do. You should be able to work the other way, too. Many problems are easily solved once they have been converted into mathematical language.

The **instructions** given in many questions must also be understood. For example, there is a marked difference between the following two questions:

Evaluate $\sqrt{5^2 \times 5^4 \div 5}$ correct to two decimal places.

and

Show that $\sqrt{5^2 \times 5^4 \div 5} = 5^{\frac{5}{2}}$

In the first question, you are asked to use your calculator to write this expression in a simpler, approximate form. The second question requires a step by step process involving index laws to show that the left hand side of the equation can be changed to look like the term on the right.

The respective solutions are

$$\sqrt{5^2 \times 5^4 \div 5} = 55 \cdot 90169944$$
$$\approx 55 \cdot 90 \quad (2 \; decimal \; places)$$

and

$$LHS = (5^{2+4-1})^{\frac{1}{2}}$$
$$= 5^{5 \times \frac{1}{2}}$$
$$= 5^{\frac{5}{2}}$$
$$= RHS$$

Here are some more questions that could be included in the HSC exam, along with an explanation of the language used and the full solution.

1. **Evaluate** $e^{2 \cdot 1}$ correct to 4 **significant figures.**

   Use your calculator and its e^x *function to obtain a decimal* ***value*** *for this number,* ***rounded off*** *correctly.*

$$e^{2 \cdot 1} = 8 \cdot 166169913$$
$$\approx 8 \cdot 166 \quad (4 \; significant \; figures)$$

2. **Solve** $|2x-1|=4$

*Use the **equation solving process** to find all possible values for* x

$$2x-1=4 \qquad \text{or} \qquad 2x-1=-4$$
$$2x=5 \qquad\qquad 2x=-3$$
$$x=2\frac{1}{2} \qquad\qquad x=-1\frac{1}{2}$$

Check: When $x=2\frac{1}{2}$,

$$\begin{aligned} LHS &= \left|2\times 2\frac{1}{2}-1\right| \\ &= |4| \\ &= 4 \\ &= RHS \end{aligned}$$

When $x=-1\frac{1}{2}$,

$$\begin{aligned} LHS &= \left|2\times -1\frac{1}{2}-1\right| \\ &= |-4| \\ &= 4 \\ &= RHS \end{aligned}$$

So, $x=2\frac{1}{2}$ and $x=-1\frac{1}{2}$ are both solutions.

3. **Find** the equation of the tangent to the curve $y=3x^2-2x$ at the point $(-1,5)$

Use an appropriate mathematical technique to find *the equation of this tangent. In this case, differentiation should be used.*

$$y = 3x^2 - 2x$$
$$\frac{dy}{dx} = 6x - 2$$

When $x = -1$,

$$\frac{dy}{dx} = 6 \times -1 - 2$$
$$= -8$$

$$y - y_1 = m(x - x_1)$$
$$y - 5 = -8(x + 1)$$
$$y - 5 = -8x - 8$$
$$y = -8x - 3$$

4. Find the **exact value** of $\sin 120^\circ$

*Use the known trigonometric ratios, and the angles of any magnitude concept, to write this number in **exact (no decimal approximation) form.***

$$\begin{aligned}\sin 120^\circ &= \sin(180-60)^\circ \\ &= \sin 60^\circ \\ &= \frac{\sqrt{3}}{2}\end{aligned}$$

5. **Calculate** the area of the square with perimeter $10cm$.

*Some **calculations** will need to be undertaken to answer this question.*

$$\begin{aligned}P &= 4s \\ 10 &= 4s \\ side &= 2\frac{1}{2}cm \\ A &= s^2 \\ &= (2\cdot 5)^2 \\ Area &= 6\cdot 25cm^2\end{aligned}$$

6. Expand and simplify $6x(x-2)-3(4-x)$

*Use the **rules of expanding** and collecting like terms to rewrite this expression in a **simpler form.***

$$\begin{aligned} 6x(x-2)-3(4-x) &= 6x^2-12x-12+3x \\ &= 6x^2-9x-12 \end{aligned}$$

7. Use $S_n=\frac{n}{2}[2a+(n-1)d]$ to **prove** that

$2+4+6+...+n=n(n+1)$

***Use the given formula** to **prove** that the left side (LHS) of the statement to be proven is equal to the right side (RHS).*

$$\begin{aligned} LHS &= \frac{n}{2}[2\times 2+(n-1)\times 2] \\ &= \frac{n}{2}[4+2n-2] \\ &= \frac{n}{2}[2n+2] \\ &= \frac{n}{2}\times 2(n+1) \\ &= n(n+1) \\ &= RHS \end{aligned}$$

8. **Express** $\frac{7}{2-\sqrt{5}}$ in the **form** $a+b\sqrt{5}$

Rewrite $\frac{7}{2-\sqrt{5}}$ *in the* ***rational denominator form*** $a+b\sqrt{5}$

$$\begin{aligned}\frac{7}{2-\sqrt{5}} &= \frac{7}{2-\sqrt{5}}\times\frac{2+\sqrt{5}}{2+\sqrt{5}} \\ &= \frac{7(2+\sqrt{5})}{2^2-(\sqrt{5})^2} \\ &= \frac{7(2+\sqrt{5})}{4-5} \\ &= \frac{7(2+\sqrt{5})}{-1} \\ &= -7(2+\sqrt{5}) \\ &= -14-7\sqrt{5}\end{aligned}$$

9. **Use** the quadratic formula to **solve** $3x^2 + x - 1 = 0$, leaving the answer in **irrational form.**

Substitute the appropriate values into the quadratic formula, and evaluate. *Do not find the square root. Just simplify the final answer using the **surd rules.***

$$\begin{aligned} x &= \frac{-b \pm \sqrt{b^2 - 4ac}}{2a} \\ &= \frac{-1 \pm \sqrt{1^2 - 4 \times 3 \times -1}}{2 \times 3} \\ &= \frac{-1 \pm \sqrt{13}}{6} \end{aligned}$$

10. **Convert** 15° to **radians**

Use the relationship $\theta^c = \theta^\circ \times \dfrac{\pi}{180}$ *to **convert degrees to radians.***

$$\begin{aligned} \theta^c &= \theta^\circ \times \frac{\pi}{180} \\ &= 15 \times \frac{\pi}{180} \\ &= \frac{\pi}{12}^c \end{aligned}$$

Hopefully, these examples have given you some idea of the specialised language used in this course.

The best way to increase your mathematical vocabulary is to work regularly with the language. You will soon realise that the same words come up quite often.

When you do meet a new word, get into the practice of writing it down somewhere, then find out what it means and write the definition in your own words.

A good knowledge of the language of mathematics is an important prerequisite for consistent success in this course.

BASIC SKILLS

Powers and Roots

IMPORTANT STUFF!

Remember those important index laws.

1. $x^m \times x^n = x^{m+n}$

eg. Simplify $a^4 \times a^5$

$$a^4 \times a^5 = a^9$$

2. $x^m \div x^n = x^{m-n}$

eg. Simplify $(2y)^{10} \div (2y)^5$

$$(2y)^{10} \div (2y)^5 = (2y)^5$$

3. $(ax^m)^n = a^n \times x^{mn}$

eg. Simplify $(3x^4)^2$

$$(3x^4)^2 = 9x^8$$

4. $x^0 = 1$

eg. Simplify $4x^0$

$4x^0 = 4$

5. $\sqrt{x} = x^{\frac{1}{2}}$

eg. Write $\sqrt{x+1}$ in index form.

$\sqrt{x+1} = (x+1)^{\frac{1}{2}}$

6. $\sqrt[n]{x^m} = x^{\frac{m}{n}}$

eg. Write $x^{\frac{2}{3}}$ in power and root form.

$x^{\frac{2}{3}} = \sqrt[3]{x^2}$

7. $ax^{-m} = \dfrac{a}{x^m}$

eg. Write $4k^{-3}$ with a positive index.

$4k^{-3} = \dfrac{4}{k^3}$

Simplifying in Algebra

IMPORTANT STUFF!

1. Only **like terms** can be added and subtracted.

eg. Simplify $3a-5b+6a$

$$\begin{aligned}3a-5b+6a &= 3a+6a-5b\\ &= 9a-5b\end{aligned}$$

2. To **multiply terms**, use the index laws.

eg. Simplify $4h\times 2h^2$

$$\begin{aligned}4h\times 2h^2 &= 4\times 2\times h\times h^2\\ &= 8h^3\end{aligned}$$

3. To **divide terms,** use the index laws or cancel common factors.

eg. Simplify $\dfrac{3p^4q}{(pq)^2}$

$$\begin{aligned}\frac{3p^4q}{(pq)^2} &= \frac{3p^4q}{p^2q^2}\\ &= \frac{3p^2}{q}\end{aligned}$$

4. When **multiplying a binomial**, make sure that both terms inside the brackets are multiplied by the outside term.

eg. Expand $6x(4-3x)$

$$\begin{aligned} 6x(4-3x) &= 6x\times 4-6x\times 3x \\ &= 24x-18x^2 \end{aligned}$$

5. To **multiply a pair of binomials**, multiply all terms in the first bracket by all terms in the second.

eg. Expand and simplify $(2y-5)(3y+1)$

$$\begin{aligned} (2y-5)(3y+1) &= 2y(3y+1)-5(3y+1) \\ &= 6y^2+2y-15y-5 \\ &= 6y^2-13y-5 \end{aligned}$$

Factorising Algebraic Expressions

IMPORTANT STUFF!

1. To **factorise,** first remove all common factors.

eg. Factorise $3e^2 f + 6e^3 f$

$$3e^2 f + 6e^3 f = 3e^2 f(1 + 2e)$$

2. To **factorise a four term expression,** factorise each pair of terms then factorise again.

eg. Factorise $5pq - 10q + 2p - 4$

$$\begin{aligned} 5pq - 10q + 2p - 4 &= 5q(p-2) + 2(p-2) \\ &= (p-2)(5p+2) \end{aligned}$$

3. To **factorise a difference of two squares,** use the rule $a^2 - b^2 = (a+b)(a-b)$

eg. Factorise $25x^2 - y^2$

$$\begin{aligned} 25x^2 - y^2 &= (5x)^2 - y^2 \\ &= (5x+y)(5x-y) \end{aligned}$$

4. A **quadratic expression** is made up of at most three terms. The **leading term**, the **middle term** and the **constant term**.

eg. $4x^2 - 2x + 1$

Leading term $= 4x^2$ Constant term $= 1$

5. The **coefficient of a term** is its number part.

eg. $4x^2 - 2x + 1$

The **leading coefficient** is 4. The coefficient of x is -2

6. To **factorise a quadratic trinomial,** find two numbers that add to give the coefficient of x and multiply to give the constant term.

eg. Factorise $c^2 + 3c - 28$

$7 + -4 = 3$ $7 \times -4 = -28$

$c^2 + 3c - 28 = (c+7)(c-4)$

7. To **factorise a general quadratic trinomial**, find two numbers that add to give the coefficient of the middle term, and multiply to give the product of the leading and constant terms. Replace the x term with two terms then factorise the four term expression.

eg. Factorise $3d^2+5d-2$

$$\begin{aligned}3d^2+5d-2&=3d^2+6d-d-2\\&=3d(d+2)-1(d+2)\\&=(d+2)(3d-1)\end{aligned}$$

8. To **factorise a difference of two cubes**, use the rule $a^3-b^3=(a-b)(a^2+ab+b^2)$.

eg. Factorise g^3-8

$$\begin{aligned}g^3-8&=g^3-2^3\\&=(g-2)(g^2+2g+4)\end{aligned}$$

9. To **factorise the sum of two cubes**, use the rule $a^3+b^3=(a+b)(a^2-ab+b^2)$

eg. Factorise $64x^3+y^6$

$$\begin{aligned}64x^3+y^6&=(4x)^3+(y^2)^3\\&=(4x+y^2)(16x^2-4xy^2+y^4)\end{aligned}$$

Algebraic Fractions

IMPORTANT STUFF!

1. To **simplify an algebraic fraction**, factorise (if possible) then cancel.

eg. Simplify $\frac{p^3 - 5p^2}{p^2q - 5pq}$

$$\frac{p^3 - 5p^2}{p^2q - 5pq} = \frac{p^2(p-5)}{pq(p-5)}$$
$$= \frac{p}{q}$$

2. To **multiply algebraic fractions**, cancel common factors then multiply numerators and denominators separately.

eg. Simplify $\frac{6}{r} \times \frac{3r^2 - r}{4}$

$$\frac{6}{r} \times \frac{3r^2 - r}{4} = \frac{6}{r} \times \frac{r(3r-1)}{4}$$
$$= \frac{3(3r-1)}{2}$$

3. To **divide algebraic fractions**, invert the second fraction and multiply.

eg. Simplify $(s^2 - 1) \div \dfrac{1}{s+1}$

$$\begin{aligned}(s^2-1) \div \frac{1}{s+1} &= (s^2-1) \times \frac{s+1}{1} \\ &= (s+1)(s-1)(s+1) \\ &= (s+1)^2(s-1)\end{aligned}$$

4. To **add or subtract algebraic fractions,** multiply the denominators then multiply diagonally to obtain the numerator.

eg. Simplify $\dfrac{2}{q-4} + \dfrac{q+2}{q}$

$$\begin{aligned}\frac{2}{q-4} + \frac{q+2}{q} &= \frac{2q + (q+2)(q-4)}{q(q-4)} \\ &= \frac{2q + q^2 - 4q + 2q - 8}{q(q-4)} \\ &= \frac{q^2 - 8}{q(q-4)}\end{aligned}$$

EXERCISE – BASIC SKILLS

(If you have trouble with any of these questions, you are advised to review the skills more thoroughly before going on. See **Top Notes HSC Maths Arithmetic & Algebra**)

1. Write $\frac{7}{x^2}$ in ax^n form.

2. Simplify $6x^2y^{-1} \times \frac{y}{4x^3}$

3. Write $(3x-1)^{\frac{1}{2}}$ in root form.

4. Write $2(a+3)^{-3}$ with a positive index.

5. Simplify $\frac{(-mn^2)^3}{2m^{-1}}$

6. Write $\frac{1}{4x-5}$ in the form $(ax+b)^n$

7. Expand and simplify $3x^2 - 4x - 2x(4-x)$

8. Expand and simplify $(3g+1)(4g-7)$

9. Expand and simplify $(2i-5)^2$

10. Expand and simplify $(y+1)^3$

11. Expand and simplify $(4k+1)(4k-1)$

12. Factorise $6x^2+4xy$ fully.

13. Factorise $a^2+9a+14$ fully.

14. Factorise $3n^2-27$ fully.

15. Factorise $4x^2-y^6$ fully.

16. Factorise $2k^2+9k-5$ fully.

17. Factorise $27m^3-1$ fully.

18. Factorise y^3+64 fully.

19. Factorise $4x^2+4xy+y^2$ fully.

20. Factorise and simplify $\dfrac{7k}{k^2-4k}$

21. Factorise and simplify $\dfrac{x^2-x-2}{x+1}$

22. Simplify $\dfrac{3m^{\frac{3}{2}}}{\frac{3}{2}}$

23. Simplify $-2\times 4(3-x)^{-3}\times -1$

24. Simplify $\dfrac{3(4g-1)^{-\frac{1}{3}}}{4\times -\frac{1}{3}}$

25. Simplify $\dfrac{x+2}{10}\times\dfrac{4}{x^2-4}$

26. Simplify $\dfrac{3z}{5}\div 2z$

27. Expand, simplify and factorise $-2(x+3)(4x)-x(x+3)^2$

28. Factorise $-2(x+3)(4x)-x(x+3)^2$ fully.

29. Simplify $\dfrac{5}{q^2+q}+\dfrac{1}{q^2-q}$

30. Simplify $\dfrac{2x(2-x)-(x^2-4)\times -1}{(2-x)^2}$

BASICS OF DIFFERENTIATION

IMPORTANT STUFF!

- To **find the limit of an algebraic expression,** substitute and evaluate, if possible.

eg. Evaluate $\lim_{x \to 0} 4x - 2$

$$\begin{aligned} \lim_{x \to 0} 4x - 2 &= 4 \times 0 - 2 \\ &= -2 \end{aligned}$$

- To **find the limit of an algebraic expression where direct substitution is impossible,** factorise and simplify, then substitute and evaluate.

eg. Find $\lim_{x \to 1} \dfrac{x^2 - x}{x - 1}$

$$\begin{aligned} \lim_{x \to 1} \frac{x^2 - x}{x - 1} &= \lim_{x \to 1} \frac{x(x-1)}{x-1} \\ &= \lim_{x \to 1} x \\ &= 1 \end{aligned}$$

- To **find the limit, as x approaches ∞, of an algebraic expression**, use the rule $\lim_{x \to \infty} \frac{1}{x} = 0$

eg. $\lim_{x \to \infty} \frac{4}{x} = 0$ **eg.** $\lim_{x \to \infty} \frac{3}{x^2} = 0$

- To **find the derivative, $\frac{dy}{dx}$ or $f'(x)$ of an algebraic expression, y or $f(x)$ by first principles**, use the formula

$$\frac{dy}{dx} = f'(x) = \lim_{h \to 0} \frac{f(x+h) - f(x)}{h}$$

eg. Differentiate $2x$ using first principles.

$$y = 2x$$

$$\begin{aligned} \frac{dy}{dx} &= \lim_{h \to 0} \frac{2(x+h) - 2x}{h} \\ &= \lim_{h \to 0} \frac{2x + 2h - 2x}{h} \\ &= \lim_{h \to 0} \frac{2h}{h} \\ &= \lim_{h \to 0} 2 \\ &= 2 \end{aligned}$$

- To **differentiate terms in the form** ax^n, use the rule

$$\frac{d}{dx}(ax^n) = anx^{n-1}$$

eg. Differentiate $3x^2$

$$\frac{d}{dx}(3x^2) = 3\times 2x^{2-1}$$
$$= 6x$$

- The **derivative of** ax **is** a, and the **derivative of any constant term is zero.**

eg. If $y = \frac{1}{2}x + 2$, find $\frac{dy}{dx}$.

$$\frac{dy}{dx} = \frac{1}{2} + 0$$
$$= \frac{1}{2}$$

- To **differentiate terms in the form** $\frac{a}{x^n}$, write them in the form ax^{-n}, then differentiate.

eg. Find $\frac{d}{dx}\left(\frac{5}{x^3}\right)$

$$\begin{aligned}\frac{d}{dx}\left(\frac{5}{x^3}\right) &= \frac{d}{dx}(5x^{-3}) \\ &= -15x^{-4} \\ &= \frac{-15}{x^4}\end{aligned}$$

- To **differentiate terms in the form** $\sqrt[n]{x^m}$, write them in the form $x^{\frac{m}{n}}$, then differentiate.

eg. Differentiate $\sqrt[4]{x^3}$ with respect to x.

$$\begin{aligned}\frac{d}{dx}(\sqrt[4]{x^3}) &= \frac{d}{dx}(x^{\frac{3}{4}}) \\ &= \frac{3}{4}x^{-\frac{1}{4}} \\ &= \frac{3}{4\sqrt[4]{x}}\end{aligned}$$

- To **differentiate a composite function,** $y = f(g(x))$, use the chain rule,

$$\frac{dy}{dx} = f'(g(x)) \times g'(x)$$

eg. If $y = (3x-1)^5$, find $\frac{dy}{dx}$.

$$\frac{dy}{dx} = 5(3x-1)^4 \times 3$$
$$= 15(3x-1)^4$$

- To **differentiate the product of two functions** $y = f(x) \times g(x)$, use the product rule,

$$\frac{dy}{dx} = f'(x) \times g(x) + f(x) \times g'(x)$$

eg. Differentiate $3x^{-1}(x^2-4)$

$$\frac{d}{dx}\left(3x^{-1}\left(x^2-4\right)\right) = -3x^{-2}(x^2-4) + 3x^{-1} \times 2x$$
$$= -3 + 12x^{-2} + 6$$
$$= 3 + \frac{12}{x^2}$$

- To **differentiate the quotient of two functions,** $y = \frac{f(x)}{g(x)}$, use the quotient rule,

$$\frac{dy}{dx} = \frac{f'(x) \times g(x) - f(x) \times g'(x)}{(g(x))^2}$$

eg. Find the derivative of $\frac{4x+3}{1-x}$

$$y = \frac{4x+3}{1-x}$$

$$\frac{dy}{dx} = \frac{4(1-x) - (4x+3) \times -1}{(1-x)^2}$$

$$= \frac{4 - 4x + 4x + 3}{(1-x)^2}$$

$$= \frac{7}{(1-x)^2}$$

- To **evaluate** $f'(a)$, differentiate $f(x)$ then replace x with the number a.

eg. If $f(x) = 3 + x^2$, find the value of $f'(-1)$

$$f'(x) = 2x$$

$$f'(-1) = 2 \times -1$$
$$= -2$$

- To **solve** $f'(x)=a$, differentiate $f(x)$, let this expression equal a, then solve the equation.

eg. Given that $g(x)=\frac{x^3}{3}+x^2-3x$, solve $g'(x)=0$

$$\frac{3x^2}{3}+2x-3=0$$

$$x^2+2x-3=0$$

$$(x+3)(x-1)=0$$

$$x=-3 \; or \; 1$$

- To **find the second derivative of an algebraic expression,** differentiate the first derivative.

eg. Find the second derivative of $-5x^{-1}$

$$y=-5x^{-1}$$

$$\frac{dy}{dx}=5x^{-2}$$

$$\frac{d^2y}{dx^2}=-10x^{-3}$$

$$=\frac{-10}{x^3}$$

EXAMPLES

1. Evaluate $\lim_{x \to 1} \frac{x}{x+1}$

Direct substitution is possible as $1+1 \neq 0$

$$\lim_{x \to 1} \frac{x}{x+1} = \frac{1}{1+1}$$
$$= \frac{1}{2}$$

2. Find $\lim_{x \to 2} \frac{x^2-4}{x-2}$

Direct substitution is impossible as $2-2=0,$ *and* 0 *cannot divide into any number. So, factorise and simplify first.*

$$\lim_{x \to 2} \frac{x^2-4}{x-2} = \lim_{x \to 2} \frac{(x+2)(x-2)}{x-2}$$
$$= \lim_{x \to 2} x+2$$

Then substitute and evaluate.

$$= 2+2$$
$$= 4$$

3. Find the limit, as x approaches ∞, of $\dfrac{3x-2}{x+1}$

Divide both terms in the numerator and both terms in the denominator by x.

$$\lim_{x\to\infty}\frac{3x-2}{x+1}=\lim_{x\to\infty}\frac{\frac{3x}{x}-\frac{2}{x}}{\frac{x}{x}+\frac{1}{x}}$$

$$=\lim_{x\to\infty}\frac{3-\frac{2}{x}}{1+\frac{1}{x}}$$

Use the rule, $\lim_{x\to\infty}\frac{1}{x}=0$, *and simplify.*

$$=\frac{3-0}{1+0}$$

$$=3$$

4. If $y=3x^2+x$, find $\dfrac{dy}{dx}$ by first principles.

Use the rule $\dfrac{dy}{dx}=\lim_{h\to 0}\dfrac{f(x+h)-f(x)}{h}$, *where* $f(x+h)$ *means* x *is replaced by* $x+h$

$$\frac{dy}{dx}=\lim_{h\to 0}\frac{3(x+h)^2+(x+h)-(3x^2+x)}{h}$$

Expand and simplify the numerator.

$$= \lim_{h \to 0} \frac{3(x^2 + 2xh + h^2) + (x + h) - (3x^2 + x)}{h}$$

$$= \lim_{h \to 0} \frac{3x^2 + 6xh + 3h^2 + x + h - 3x^2 - x}{h}$$

$$= \lim_{h \to 0} \frac{6xh + 3h^2 + h}{h}$$

Factorise and simplify.

$$= \lim_{h \to 0} \frac{h(6x + 3h + 1)}{h}$$

$$= \lim_{h \to 0} 6x + 3h + 1$$

Substitute and evaluate.

$$= 6x + 0 + 1$$
$$= 6x + 1$$

5. Differentiate $4x^3 + 2x^{-1}$

Use the rule $\frac{d}{dx}(ax^n) = anx^{n-1}$ *to differentiate both terms.*

$$\begin{aligned}\frac{d}{dx}(4x^3 + 2x^{-1}) &= 4\times 3x^{3-1} + 2\times -1x^{-1-1} \\ &= 12x^2 - 2x^{-2}\end{aligned}$$

6. If $f(x) = 3x - 5$, write an expression for $f'(x)$

Remember that $\frac{d}{dx}(ax) = a$ *and* $\frac{d}{dx}(a) = 0$

$$\begin{aligned}f'(x) &= 3 - 0 \\ &= 3\end{aligned}$$

7. Find the derivative of $\frac{7}{x^2}$

Use the index laws to change the fraction to a product.

$$\frac{d}{dx}\left(\frac{7}{x^2}\right) = \frac{d}{dx}\left(7x^{-2}\right)$$

Differentiate then convert to a fraction with a positive index.

$$= -14x^{-3}$$
$$= \frac{-14}{x^3}$$

8. Find $\frac{d}{dx}(\sqrt[4]{x^5})$

Use the index law, $\sqrt[n]{x^m} = x^{\frac{m}{n}}$ *to rewrite the question.*

$$\frac{d}{dx}(\sqrt[4]{x^5}) = \frac{d}{dx}(x^{\frac{5}{4}})$$

Differentiate then change back to power and root form.

$$= \frac{5}{4}(x^{\frac{5}{4}-\frac{4}{4}})$$
$$= \frac{5}{4}(x^{\frac{1}{4}})$$
$$= \frac{5\sqrt[4]{x}}{4}$$

9. $y=(2x-1)^4,\ \frac{dy}{dx}=?$

Use the chain rule, $\frac{dy}{dx}=f'(g(x))\times g'(x)$ *and simplify.*

$$\begin{aligned}\frac{dy}{dx}&=4(2x-1)^3\times 2\\&=8(2x-1)^3\end{aligned}$$

10. Use the product rule to differentiate $4x^3(x-2)$, with respect to x.

Use the product rule, $\frac{dy}{dx}=f'(x)\times g(x)+f(x)\times g'(x)$

$$\begin{aligned}\frac{dy}{dx}&=12x^2(x-2)+1\times 4x^3\\&=12x^3-24x^2+4x^3\\&=16x^3-24x^2\end{aligned}$$

11. Differentiate $x^2\sqrt{1+x}$

Use the product rule, chain rule and index law,

$\sqrt{m} = m^{\frac{1}{2}}$ *together.*

$$\frac{d}{dx}(x^2(1+x)^{\frac{1}{2}}) = 2x \times (1+x)^{\frac{1}{2}} + x^2 \times \frac{1}{2}(1+x)^{-\frac{1}{2}}$$

Simplify, changing terms back to power and root form.

$$= 2x\sqrt{1+x} + \frac{x^2}{2\sqrt{1+x}}$$

12. Given that $p(x) = \dfrac{4x^2 - 1}{4x - 1}$, find $p'(x)$.

Use the quotient rule, $\dfrac{dy}{dx} = \dfrac{f'(x) \times g(x) - f(x) \times g'(x)}{(g(x))^2}$

$$p'(x) = \frac{8x(4x-1) - (4x^2 - 1)4}{(4x-1)^2}$$

Expand and simplify.

$$= \frac{32x^2 - 8x - 16x^2 + 4}{(4x-1)^2}$$
$$= \frac{16x^2 - 8x + 4}{(4x-1)^2}$$

Factorise, and cancel, if possible.

$$= \frac{4(4x^2 - 2x + 1)}{(4x-1)^2}$$

13. If $f(x) = x^3 + 7x - 3$, find the value of $f(2) + f'(0)$

Find $f(2)$ by substitution.

$$f(2) = 2^3 + 7 \times 2 - 3$$
$$= 19$$

Find $f'(x)$ then $f'(0)$

$$f'(x) = 3x^2 + 7$$

$$f'(0) = 3 \times 0^2 + 7$$
$$= 7$$

Add these two results.

$$\begin{aligned} f(2) + f'(0) &= 19 + 7 \\ &= 26 \end{aligned}$$

14. If $g(x) = x^3 - \frac{5x^2}{2} + 2x$, solve $g'(x) = 0$

Find $g'(x)$ *and allow it to equal zero.*

$$g'(x) = 3x^2 - 5x + 2$$

$$3x^2 - 5x + 2 = 0$$

Solve by factorising the left side of the equation first.

$$\begin{aligned} 3x^2 - 3x - 2x + 2 &= 0 \\ 3x(x-1) - 2(x-1) &= 0 \\ (x-1)(3x-2) &= 0 \end{aligned}$$

$x = 1$ *or* $\frac{2}{3}$

15. Find the second derivative of $\dfrac{1}{(x+1)^2}$

Find the first derivative by first changing to $(ax+b)^n$ *form.*

$$y = \frac{1}{(x+1)^2}$$
$$= (x+1)^{-2}$$

$$\frac{dy}{dx} = -2(x+1)^{-3}$$

Differentiate the answer and change back to power and root form.

$$\frac{d^2y}{dx^2} = 6(x+1)^{-4}$$
$$= \frac{6}{(x+1)^4}$$

EXERCISE – BASICS OF DIFFERENTIATION

1. Find $\lim_{x \to 0} \frac{x+2}{x-1}$

2. Evaluate $\lim_{x \to 1} \frac{x^2-4x+3}{x-1}$

3. Find the limit as x approaches ∞ of $\frac{x-3}{x^2+2}$

4. Find $\lim_{h \to 0} \frac{5(x+h)-2-(5x-2)}{h}$

5. Differentiate $3x+1$ by first principles.

6. Find $\frac{d}{dx}(3x^5 - x^{-3} + 4x)$

7. Find the first derivative of $\frac{4}{x}$

8. If $y = \frac{2\sqrt{x}}{3}$, find $\frac{dy}{dx}$

9. If $f(x) = x^3 - 4x^2 + 6x + 1$, find

(a) $f'(x)$

(b) $f'(2)$

(c) $f'(-1)$

(d) $f(0) + f'(0)$

10. Expand then differentiate $x(x-1)$

11. Expand then differentiate $(2x-5)^2$

12. Simplify then differentiate $\dfrac{3x^2 + x}{x}$

13. Differentiate $(4-3x)^7$

14. Use the product rule to differentiate $(x+3)(6x-1)$

15. Use the quotient rule to differentiate $\dfrac{x+3}{6x-1}$

16. If $f(x) = \sqrt[3]{2x-1}$, find $f'(x)$

17. Find the derivative of $\dfrac{1}{x+2}$

18. If $g(x) = x^2 + 8x + 12$, find the value of $g'(-3)$

19. Find the second derivative of $5x^2 - 3x + 2$

20. If $y = 2(4x^2 - 1)^3$, find

(a) $\dfrac{dy}{dx}$

(b) The values of x when $\dfrac{dy}{dx} = 0$

(c) $\dfrac{d^2y}{dx^2}$

(d) The values of x when $\dfrac{d^2y}{dx^2} = 0$

CHALLENGE EXERCISE – BASICS OF DIFFERENTIATION

1. Show that $\frac{\sqrt{x}-2}{x-4}=\frac{\sqrt{x}-2}{x-4}\times\frac{\sqrt{x}+2}{\sqrt{x}+2}=\frac{1}{\sqrt{x}+2}$

Hence find $\lim \frac{\sqrt{x}-2}{x-4}$

2. If $f(x)=ax^3+bx^2-2x+1, f(-1)=2$ and $f'(-1)=0$, find the values of a and b

3. If $y=(x^2+1)(x^3-x+2)$, find $\frac{dy}{dx}$

4. Differentiate $(ax+b)^{k-1}$ with respect to x

5. Find the derivative of $\frac{1}{2\sqrt{x}}$

6. Differentiate $y=x^2+x-2$ from first principles.

7. If $y=x^n$, show that $\frac{dy}{dx}=\frac{ny}{x}$

8. Show that $\dfrac{\sqrt{x+h}-\sqrt{x}}{h}=\dfrac{1}{\sqrt{x+h}+\sqrt{x}}$. Hence, differentiate $\sqrt{x}$ from first principles.

9. Differentiate $\dfrac{(x-3)^2}{3-x}$

10. For which value of x is the derivative of $\dfrac{2x^2+x-6}{x^2+2x}$ equal to zero?

BASICS OF INTEGRATION

IMPORTANT STUFF!

- The opposite operation of differentiation is called **anti-differentiation**, or **integration**. This process gives the **primitive function**, or **indefinite integral**.

- The **primitive** of $\frac{dy}{dx} = ax^n$ is given by

$y = \frac{ax^{n+1}}{n+1} + c$, where c is a constant.

eg. Find the primitive of $2x^3$

$$\frac{dy}{dx} = 2x^3$$

$$y = \frac{2x^4}{4} + c$$
$$= \frac{x^4}{2} + c$$

eg. Integrate $x - 4x^2$

$$f'(x) = x - 4x^2$$

$$f(x) = \frac{x^2}{2} - \frac{4x^3}{3} + c$$

- This can be written in **integral form** as $\int ax^n dx = \frac{ax^{n+1}}{n+1} + c$

eg. Find $\int 2x^3\, dx$

$$\int 2x^3\, dx = \frac{2x^4}{4} + c$$
$$= \frac{x^4}{2} + c$$

- The **integral of any number,** a is the product of that number and x, plus a constant. That is, $\int a\, dx = ax + c$

eg. Find the integral of 4.

$$\int 4\, dx = 4x + c$$

- Another type of expression that can be integrated is $(ax+b)^n$. Use the rule,

$$\int (ax+b)^n\, dx = \frac{(ax+b)^{n+1}}{a(n+1)} + c$$

eg. $\int (4x+3)^5\, dx = ?$

$$\int (4x+3)^5\, dx = \frac{(4x+3)^6}{4\times 6} + c$$
$$= \frac{(4x+3)^6}{24} + c$$

- To **find the value of the constant** c, integrate, substitute then solve for c.

eg. If $\frac{dy}{dx} = x^{-2}$, and $y = 3$ when $x = 1$, find the value of the constant of integration.

$$y = \frac{x^{-1}}{-1} + c$$
$$y = -\frac{1}{x} + c$$

$$3 = -1 + c$$
$$c = 4$$

The constant of integration is 4

- The **definite integral of a function** is found by integrating, substituting and subtracting. No constant of integration is required.

$$\int_a^b x^n dx = \left[\frac{x^{n+1}}{n+1}\right]_a^b$$
$$= \frac{b^{n+1}}{n+1} - \frac{a^{n+1}}{n+1}$$

eg. Evaluate $\int_0^1 7x\,dx$

$$\int_0^1 7x\,dx = \left[\frac{7x^2}{2}\right]_0^1$$
$$= \frac{7}{2} - 0$$
$$= \frac{7}{2}$$

EXAMPLES

1. Integrate $5x^3$

Integrate by using the rule $\int ax^n dx = \frac{ax^{n+1}}{n+1} + c$

$$\int 5x^3\,dx = \frac{5x^{3+1}}{3+1} + c$$
$$= \frac{5x^4}{4} + c$$

2. Find the primitive of $6x+1$

Integrate both terms separately, remembering that the integral of a *is* ax*.*

$$\int 6x+1\ \,dx = \int 6x\ \,dx + \int 1\ \,dx$$
$$= \frac{6x^2}{2} + x + c$$

Simplify.

$$= 3x^2 + x + c$$

3. If $\frac{dy}{dx}=\frac{4}{x^2}$, express y in terms of x

Change to the form ax^n before integrating.

$$\frac{dy}{dx}=4x^{-2}$$

Integrate then change back to positive index form.

$$\begin{aligned} y &= \int 4x^{-2}\,dx \\ &= \frac{4x^{-1}}{-1}+c \\ &= \frac{-4}{x}+c \end{aligned}$$

4. $\int (3x+7)^5\,dx = ?$

Integrate using the rule, $\int (ax+b)^n\,dx=\frac{(ax+b)^{n+1}}{a(n+1)}+c$

$$\begin{aligned} \int (3x+7)^5\,dx &= \frac{(3x+7)^6}{3\times 6}+c \\ &= \frac{(3x+7)^6}{18}+c \end{aligned}$$

5. Integrate $(3x^2-2)^2$ with respect to x.

The expression is not in the form $(ax+b)^n$ so it can only be integrated by expanding and simplifying first.

$$\begin{aligned}\int(3x^2+2)^2\,dx &= \int((3x^2)^2+2\times3x^2\times2+2^2)\,dx\\ &= \int(9x^4+12x^2+4)\,dx\end{aligned}$$

Integrate and simplify.

$$\begin{aligned}&=\frac{9x^5}{5}+\frac{12x^3}{3}+4x+c\\ &=\frac{9x^5}{5}+4x^3+4x+c\end{aligned}$$

6. Given that $f'(x)=\dfrac{3x^{-\frac{1}{2}}}{8}$ and $f(4)=1$, write $f(x)$ in terms of x.

Integrate to find $f(x)$.

$$\begin{aligned}f(x)&=\frac{3x^{\frac{1}{2}}}{\frac{1}{2}\times8}+c\\ &=\frac{3x^{\frac{1}{2}}}{4}+c\\ &=\frac{3\sqrt{x}}{4}+c\end{aligned}$$

Substitute and solve for c.

$$f(4)=\frac{3\times\sqrt{4}}{4}+c=1$$

$$1\frac{1}{2}+c=1$$

$$c=-\frac{1}{2}$$

Substitute $c=-\frac{1}{2}$ *into* $f(x)$.

$$f(x)=\frac{3\sqrt{x}}{4}-\frac{1}{2}$$

7. Evaluate $\int_{-1}^{2}\frac{x^3-2}{x^3}\,dx$

Write $\frac{x^3-2}{x^3}$ *as two terms in the form* ax^n*, then integrate and simplify.*

$$\int_{-1}^{2}\frac{x^3-2}{x^3}\,dx=\int_{-1}^{2}(1-2x^{-3})\,dx$$

$$=\left[x-\frac{2x^{-2}}{-2}\right]_{-1}^{2}$$

$$=\left[x+\frac{1}{x^2}\right]_{-1}^{2}$$

Substitute and subtract.

$$= \left[\left(2 + \frac{1}{2^2}\right) - \left(-1 + \frac{1}{(-1)^2}\right)\right]$$

$$= 2\frac{1}{4} - 0$$

$$= 2\frac{1}{4}$$

EXERCISE – BASICS OF INTEGRATION

1. Find $\int 4\,dx$

2. Evaluate $\int_0^2 x\,dx$

3. Find the primitive of $3x^2$

4. Evaluate $\int_{-1}^{1} \frac{x^2}{2} - x^3\,dx$

5. If $\frac{dy}{dx} = \frac{5}{x^2}$, write y in terms of x.

6. $\int 2\sqrt{x}\,dx = ?$

7. Find the value of $\int_0^3 dx$

8. Find $3\int x^{-3}\,dx$

9. Integrate $\frac{1}{\sqrt[3]{x^2}}$ with respect to x.

10. Integrate $(y+1)^4$ with respect to y.

11. Evaluate $\int_0^1 (3x-10)^{-2}\,dx$

12. Find $\int \frac{dx}{(1-2x)^3}$

13. Show that $\frac{x-1}{1-x}=1$ and hence find $\int \frac{x-1}{1-x}\,dx$

14. Factorise and simplify $\frac{4x^2-4x+1}{4x-2}$ to find $\int \frac{4x^2-4x+1}{4x-2}\,dx$

15. Evaluate $\int_{-1}^{1} \frac{x^5+x^3-x}{x^3}\,dx$

16. If $f'(x)=9\sqrt{x}+2$ and $f(2)=0$, find the value of $f(1)$.

17. If $\frac{dy}{dx}=(x-2)(x+2)$ and $y=2$ when $x=2$, write an expression for y in terms of x.

18. If $\frac{d^2y}{dx^2}=-1$ and $\frac{dy}{dx}=y=-1$ when $x=-1$, write an expression for y in terms of x.

19. Show that the integral of $\sqrt{1+4x}$ is $\frac{\sqrt{(1+4x)^3}}{6}+c$

20. Prove that $\int_{-2}^{2} x^4+3x^2\,dx = 2\int_{0}^{2} x^4+3x^2\,dx$

CHALLENGE EXERCISE – BASICS OF INTEGRATION

1. Find $\int (x^2+3)^2\,dx$

2. If $g'(x)=2x(x^2+3)^3$, write an expression for $g(x)$.

3. Evaluate $\int_1^4 x\sqrt{x}\,dx$.

4. Show that if $y=\sqrt{x^2-1}$, $\frac{dy}{dx}=\frac{x}{\sqrt{x^2-1}}$. Hence or otherwise, find the exact value of $\int_2^3 \frac{x}{\sqrt{x^2-1}}\,dx$.

5. Evaluate $\pi\int_1^3 (x^4+4x^{-2})\,dx$, leaving the answer in terms of π.

6. If $y=\sqrt[3]{x}$, evaluate $\int_1^8 x\,dy$.

7. Find $\frac{d}{dx}\left(\frac{x+1}{x-1}\right)$. Hence, find $\int \frac{dx}{(x-1)^2}$.

8. Evaluate $\left|\int_{-1}^{0} x(x+1)(x-2)\,dx\right|$.

9. If $\frac{ds}{dt}=\frac{(2t^2-7t-15)(2t+3)}{t-5}$, find its primitive function.

10. Show that

$$\int_{-1}^{0} 2(x^2+3x-1)\,dx + \int_{0}^{2} 2(x^2+3x-1)\,dx = \int_{-1}^{2} 2(x^2+3x-1)\,dx.$$

SOLUTIONS

BASIC SKILLS

Question 1

$$\frac{7}{x^2} = 7x^{-2}$$

Question 2

$$6x^2y^{-1} \times \frac{y}{4x^3} = \frac{6x^2y^0}{4x^3}$$
$$= \frac{3}{2x}$$

Question 3

$$(3x-1)^{\frac{1}{2}} = \sqrt{3x-1}$$

Question 4

$$2(a+3)^{-3} = 2 \times \frac{1}{(a+3)^3}$$
$$= \frac{2}{(a+3)^3}$$

Question 5

$$\frac{(-mn^2)^3}{2m^{-1}} = \frac{-m^3n^6}{2m^{-1}}$$
$$= \frac{-m^4n^6}{2}$$

Question 6

$$\frac{1}{4x-5} = (4x-5)^{-1}$$

Question 7

$$3x^2 - 4x - 2x(4-x)$$
$$= 3x^2 - 4x - 8x + 2x^2$$
$$= 5x^2 - 12x$$

Question 8

$$(3g+1)(4g-7)$$
$$= 3g(4g-7) + 1(4g-7)$$
$$= 12g^2 - 21g + 4g - 7$$
$$= 12g^2 - 17g - 7$$

Question 9

$$(2i-5)^2 = (2i-5)(2i-5)$$
$$= 4i^2 - 10i - 10i + 25$$
$$= 4i^2 - 20i + 25$$

Question 10

$$(y+1)^3$$
$$= (y+1)(y+1)^2$$
$$= (y+1)(y^2+2y+1)$$
$$= y^3 + 2y^2 + y + y^2 + 2y + 1$$
$$= y^3 + 3y^2 + 3y + 1$$

Question 11

$$(4k+1)(4k-1) = (4k)^2 - 1$$
$$= 16k^2 - 1$$

Question 12

$$6x^2 + 4xy = 2x(3x+2y)$$

Question 13

$$a^2 + 9a + 14 = (a+7)(a+2)$$

Question 14

$$3n^2 - 27 = 3(n^2 - 9)$$
$$= 3(n+3)(n-3)$$

Question 15

$$4x^2 - y^6 = (2x)^2 - (y^3)^2$$
$$= (2x + y^3)(2x - y^3)$$

Question 16

$$2k^2 + 9k - 5$$
$$= 2k^2 + 10k - k - 5$$
$$= 2k(k+5) - 1(k+5)$$
$$= (k+5)(2k-1)$$

Question 17

$$27m^3 - 1$$
$$= (3m)^3 - 1^3$$
$$= (3m-1)((3m)^2 + 3m \times 1 + 1^2)$$
$$= (3m-1)(9m^2 + 3m + 1)$$

Question 18

$$y^3 + 64 = y^3 + 4^3$$
$$= (y+4)(y^2 - 4y + 16)$$

Question 19

$$4x^2 + 4xy + y^2$$
$$= (2x)^2 + 2xy + 2xy + y^2$$
$$= 2x(2x + y) + y(2x + y)$$
$$= (2x + y)(2x + y)$$
$$= (2x + y)^2$$

Question 20

$$\frac{7k}{k^2 - 4k} = \frac{7k}{k(k-4)}$$
$$= \frac{7}{k-4}$$

Question 21

$$\frac{x^2 - x - 2}{x+1} = \frac{(x-2)(x+1)}{x+1}$$
$$= x - 2$$

Question 22

$$\frac{3m^{\frac{3}{2}}}{\frac{3}{2}} = \frac{2}{3} \times 3\sqrt{m^3}$$
$$= 2\sqrt{m^3}$$

Question 23

$$-2 \times 4(3-x)^{-3} \times -1 = \frac{-2 \times 4 \times -1}{(3-x)^3}$$
$$= \frac{8}{(3-x)^3}$$

Question 24

$$\frac{3(4g-1)^{-\frac{1}{3}}}{4 \times -\frac{1}{3}} = \frac{3}{-\frac{4}{3}\sqrt[3]{4g-1}}$$
$$= -\frac{3}{4} \times \frac{3}{\sqrt[3]{4g-1}}$$
$$= \frac{-9}{4\sqrt[3]{4g-1}}$$

Question 25

$$\frac{x+2}{10}\times\frac{4}{x^2-4}=\frac{2(x+2)}{5(x+2)(x-2)}$$
$$=\frac{2}{5(x-2)}$$

Question 26

$$\frac{3z}{5}\div 2z=\frac{3z}{5}\times\frac{1}{2z}$$
$$=\frac{3}{10}$$

Question 27

$$-2(x+3)(4x)-x(x+3)^2$$
$$=-8x(x+3)-x(x^2+6x+9)$$
$$=-8x^2-24x-x^3-6x^2-9x$$
$$=-x^3-14x^2-33x$$
$$=-x(x^2+14x+33)$$
$$=-x(x+11)(x+3)$$

Question 28

$$-2(x+3)(4x)-x(x+3)^2$$
$$=(x+3)(-8x-x(x+3))$$
$$=(x+3)(-x^2-11x)$$
$$=-x(x+3)(x+11)$$

Question 29

$$\frac{5}{q^2+q}+\frac{1}{q^2-q}=\frac{5}{q(q+1)}+\frac{1}{q(q-1)}$$
$$=\frac{5(q-1)+1(q+1)}{q(q+1)(q-1)}$$
$$=\frac{5q-5+q+1}{q(q+1)(q-1)}$$
$$=\frac{6q-4}{q(q+1)(q-1)}$$
$$=\frac{2(3q-2)}{q(q+1)(q-1)}$$

Question 30

$$\frac{2x(2-x)-(x^2-4)\times-1}{(2-x)^2}$$
$$=\frac{4x-2x^2+x^2-4}{(2-x)^2}$$
$$=\frac{-x^2+4x-4}{(2-x)^2}$$
$$=\frac{-(x^2-4x+4)}{(2-x)^2}$$
$$=\frac{-(x-2)^2}{(2-x)^2}$$
$$=\frac{-(2-x)^2}{(2-x)^2}$$
$$=-1$$

BASICS OF DIFFERENTIATION

Question 1

$$\lim_{x \to 0} \frac{x+2}{x-1} = \frac{0+2}{0-1}$$
$$= -2$$

Question 2

$$\lim_{x \to 1} \frac{(x-3)(x-1)}{x-1} = \lim_{x \to 1} x-3$$
$$= 1-3$$
$$= -2$$

Question 3

$$\lim_{x \to \infty} \frac{x-3}{x^2+2} = \lim_{x \to \infty} \frac{\frac{x}{x^2} - \frac{3}{x^2}}{\frac{x^2}{x^2} + \frac{2}{x^2}}$$
$$= \lim_{x \to \infty} \frac{\frac{1}{x} - \frac{3}{x^2}}{1 + \frac{2}{x^2}}$$
$$= \frac{0-0}{1+0}$$
$$= 0$$

Question 4

$$\lim_{h \to 0} \frac{5(x+h) - 2 - (5x-2)}{h}$$
$$= \lim_{h \to 0} \frac{5x + 5h - 2 - 5x + 2}{h}$$
$$= \lim_{h \to 0} \frac{5h}{h}$$
$$= \lim_{h \to 0} 5$$
$$= 5$$

Question 5

$$f(x) = 3x+1$$

$$f'(x) = \lim_{h \to 0} \frac{3(x+h) + 1 - (3x+1)}{h}$$

$$= \lim_{h \to 0} \frac{3x + 3h + 1 - 3x - 1}{h}$$

$$= \lim_{h \to 0} \frac{3h}{h}$$
$$= 3$$

Question 6

$$\frac{d}{dx}(3x^5 - x^{-3} + 4x)$$
$$= 15x^4 + 3x^{-4} + 4$$

Question 7

$$\frac{d}{dx}\left(\frac{4}{x}\right)=\frac{d}{dx}\left(4x^{-1}\right)$$
$$=-4x^{-2}$$
$$=\frac{-4}{x^2}$$

Question 8

$$y=\frac{2\sqrt{x}}{3}=\frac{2}{3}x^{\frac{1}{2}}$$

$$\frac{dy}{dx}=\frac{2}{3}\times\frac{1}{2}x^{\frac{1}{2}-1}$$
$$=\frac{1}{3}x^{-\frac{1}{2}}$$
$$=\frac{1}{3\sqrt{x}}$$

Question 9

(a) $f'(x)=3x^2-8x+6$

(b) $f'(2)=12-16+6=2$

(c) $f'(-1)=3+8+6=17$

(d) $f(0)-f'(0)=1-6=-5$

Question 10

$$\frac{d}{dx}(x(x-1))=\frac{d}{dx}(x^2-x)$$
$$=2x-1$$

Question 11

$$\frac{d}{dx}((2x-5)^2)=\frac{d}{dx}(4x^2-20x+25)$$
$$=8x-20$$

Question 12

$$y=\frac{3x^2}{x}+\frac{x}{x}$$
$$=3x+1$$

$$\frac{dy}{dx}=3$$

Question 13

$$\frac{d}{dx}((4-3x)^7)=7(4-3x)^6\times-3$$
$$=-21(4-3x)^6$$

Question 14

$$y = (x+3)(6x-1)$$

$$\begin{aligned}\frac{dy}{dx} &= 1(6x-1) + 6(x+3)\\ &= 6x - 1 + 6x + 18\\ &= 12x + 17\end{aligned}$$

Question 15

$$f(x) = \frac{x+3}{6x-1}$$

$$\begin{aligned}f'(x) &= \frac{1(6x-1) - 6(x+3)}{(6x-1)^2}\\ &= \frac{6x-1-6x-18}{(6x-1)^2}\\ &= \frac{-19}{(6x-1)^2}\end{aligned}$$

Question 16

$$\begin{aligned}f(x) &= \sqrt[3]{2x-1}\\ &= (2x-1)^{\frac{1}{3}}\end{aligned}$$

$$\begin{aligned}f'(x) &= \frac{1}{3}(2x-1)^{-\frac{2}{3}} \times 2\\ &= \frac{2}{3(2x-1)^{\frac{2}{3}}}\\ &= \frac{2}{3\sqrt[3]{(2x-1)^2}}\end{aligned}$$

Question 17

$$\begin{aligned}y &= \frac{1}{x+2}\\ &= (x+2)^{-1}\end{aligned}$$

$$\begin{aligned}\frac{dy}{dx} &= -(x+2)^{-2}\\ &= \frac{-1}{(x+2)^2}\end{aligned}$$

Question 18

$$g'(x) = 2x + 8$$

$$\begin{aligned}g'(-3) &= 2 \times -3 + 8\\ &= 2\end{aligned}$$

Question 19

$$y = 5x^2 - 3x + 2$$

$$\frac{dy}{dx} = 10x - 3$$

$$\frac{d^2y}{dx^2} = 10$$

Question 20

$$y = 2(4x^2 - 1)^3$$

$$\frac{dy}{dx} = 6(4x^2 - 1)^2 \times 8x$$

$$= 48x(4x^2 - 1)^2$$

$$48x(4x^2 - 1)^2 = 0$$

$$48x(2x + 1)(2x - 1) = 0$$

$$x = 0, -\frac{1}{2} \ or \ \frac{1}{2}$$

$$y'' = 48(4x^2 - 1)^2 +$$

$$48x \times 2(4x^2 - 1) \times 8x$$

$$= 48(4x^2 - 1)(4x^2 - 1 + 16x^2)$$

$$= 48(2x + 1)(2x - 1)(20x^2 - 1)$$

$$48(2x + 1)(2x - 1)(20x^2 - 1) = 0$$

$$x = \pm\frac{1}{2} \text{ or } \pm\frac{1}{\sqrt{20}}$$

BASICS OF DIFFERENTIATION (CHALLENGE)

Question 1

$$\frac{\sqrt{x}-2}{x-4}=\frac{\sqrt{x}-2}{x-4}\times\frac{\sqrt{x}+2}{\sqrt{x}+2}$$

$$=\frac{(\sqrt{x})^2+2\sqrt{x}-2\sqrt{x}-2^2}{(x-4)(\sqrt{x}+2)}$$

$$=\frac{x-4}{(x-4)(\sqrt{x}+2)}$$

$$=\frac{1}{\sqrt{x}+2}$$

$$\lim_{x\to 4}\frac{\sqrt{x}-2}{x-4}=\lim_{x\to 4}\frac{1}{\sqrt{x}+2}$$

$$=\frac{1}{\sqrt{4}+2}$$

$$=\frac{1}{4}$$

Question 2

$$f(x)=ax^3+bx^2-2x+1$$

$$f(-1)=a\times-1+b\times1-2\times-1+1$$

$$=-a+b+2+1$$

$$=-a+b+3$$

$$f(-1)=2$$

$$-a+b+3=2$$

$$a=b+1$$

$$f'(x)=3ax^2+2bx-2$$

$$f'(-1)=3a-2b-2$$

$$f'(-1)=0$$

$$3a-2b-2=0$$

Solving these two equations simultaneously, substitute $a=b+1$ *into* $3a-2b-2=0$

$$3(b+1)-2b-2=0$$

$$3b+3-2b-2=0$$

$$b=-1$$

$$a=b+1$$

$$a=-1+1$$

$$a=0$$

Checking these solutions by substitution.

$$3a - 2b - 2 = 3\times 0 - 2\times -1 - 2$$
$$= 0 + 2 - 2$$
$$= 0$$

So, $a = 0$ *and* $b = -1$

Question 3

$$y = (x^2 + 1)(x^3 - x + 2)$$
$$= f(x)\times g(x)$$

$$\frac{dy}{dx} = f'(x)\times g(x) + g'(x)\times f(x)$$
$$= 2x(x^3 - x + 2) + (3x^2 - 1)(x^2 + 1)$$
$$= 2x^4 - 2x^2 + 4x + 3x^4 + 3x^2 - x^2 - 1$$
$$= 5x^4 + 4x - 1$$

Question 4

$$y = (ax + b)^{k-1}$$

$$\frac{dy}{dx} = (k-1)(ax+b)^{k-2}\times a$$
$$= a(k-1)(ax+b)^{k-2}$$

Question 5

$$y = \frac{1}{2\sqrt{x}}$$
$$= \frac{1}{2}x^{-\frac{1}{2}}$$

$$\frac{dy}{dx} = -\frac{1}{4}x^{-\frac{3}{2}}$$
$$= -\frac{1}{4\sqrt{x^3}}$$

Question 6

$$\frac{dy}{dx} = \lim_{h\to 0}\frac{f(x+h) - f(x)}{h}$$

$$= \lim_{h\to 0}$$
$$\frac{x^2 + 2xh + h^2 + x + h - 2 - x^2 - x +}{h}$$

$$= \lim_{h\to 0}\frac{2xh + h^2 + h}{h}$$

$$= \lim_{h\to 0}\frac{h(2x + h + 1)}{h}$$

$$= \lim_{h\to 0} 2x + h + 1$$

$$= 2x + 0 + 1$$
$$= 2x + 1$$

Question 7

$$y = x^n$$

$$\frac{dy}{dx} = nx^{n-1}$$
$$= nx^n \times x^{-1}$$
$$= \frac{nx^n}{x}$$
$$= \frac{ny}{x}$$

$$y = \sqrt{x}$$

$$\frac{dy}{dx} = \lim_{h \to 0} \frac{\sqrt{x+h} - \sqrt{x}}{h}$$
$$= \lim_{h \to 0} \frac{1}{\sqrt{x+h} + \sqrt{x}}$$

$$= \frac{1}{\sqrt{x+0} + \sqrt{x}}$$
$$= \frac{1}{2\sqrt{x}}$$

Question 8

$$LHS = \frac{\sqrt{x+h} - \sqrt{x}}{h} \times \frac{\sqrt{x+h} + \sqrt{x}}{\sqrt{x+h} + \sqrt{x}}$$
$$= \frac{(\sqrt{x+h})^2 - (\sqrt{x})^2}{h(\sqrt{x+h} + \sqrt{x})}$$
$$= \frac{x+h-x}{h(\sqrt{x+h} + \sqrt{x})}$$
$$= \frac{h}{h(\sqrt{x+h} + \sqrt{x})}$$
$$= \frac{1}{\sqrt{x+h} + \sqrt{x}}$$
$$= RHS$$

Question 9

$$y = \frac{(x-3)^2}{3-x}$$
$$= \frac{(x-3)^2}{-(x-3)}$$
$$= -(x-3)$$
$$= -x+3$$

$$\frac{dy}{dx} = -1$$

Question 10

$$f(x) = \frac{2x^2 + x - 6}{x^2 + 2x}$$

$$= \frac{2x^2 + 4x - 3x - 6}{x(x + 2)}$$

$$= \frac{2x(x + 2) - 3(x + 2)}{x(x + 2)}$$

$$= \frac{(x + 2)(2x - 3)}{x(x + 2)}$$

$$= \frac{2x - 3}{x}$$

$$= 2 - \frac{3}{x}$$

$$= 2 - 3x^{-1}$$

$$f'(x) = 3x^{-2}$$

$$= \frac{3}{x^2}$$

$\frac{3}{x^2}$ *cannot equal zero, so there are no possible values of* x.

BASICS OF INTEGRATION

Question 1

$$\int 4\,dx = 4x + c$$

Question 2

$$\int_0^2 x\,dx = \left[\frac{x^2}{2}\right]_0^2$$
$$= \left[\frac{2^2}{2} - \frac{0^2}{2}\right]$$
$$= 2$$

Question 3

$$\frac{dy}{dx} = 3x^2$$

$$y = \frac{3x^3}{3} + c$$
$$= x^3 + c$$

Question 4

$$\int_{-1}^{1} \frac{x^2}{2} - x^3\,dx = \left[\frac{x^3}{6} - \frac{x^4}{4}\right]_{-1}^{1}$$
$$= \left[\left(\frac{1}{6} - \frac{1}{4}\right) - \left(\frac{-1}{6} - \frac{1}{4}\right)\right]$$
$$= -\frac{1}{12} - \left(-\frac{5}{12}\right)$$
$$= \frac{1}{3}$$

Question 5

$$\frac{dy}{dx} = \frac{5}{x^2}$$
$$= 5x^{-2}$$

$$y = \frac{5x^{-1}}{-1} + c$$
$$= -\frac{5}{x} + c$$

Question 6

$$\int 2\sqrt{x}\,dx = \int 2x^{\frac{1}{2}}\,dx$$
$$= \frac{2x^{\frac{3}{2}}}{\frac{3}{2}} + c$$
$$= \frac{4\sqrt{x^3}}{3} + c$$

Question 7

$$\int_0^3 dx = \int_0^3 1\,dx$$
$$= [x]_0^3$$
$$= 3 - 0$$
$$= 3$$

Question 8

$$3\int x^{-3}\,dx = \frac{3x^{-2}}{-2} + c$$
$$= -\frac{3}{2x^2} + c$$

Question 9

$$\frac{dy}{dx} = \frac{1}{\sqrt[3]{x^2}}$$
$$= \frac{1}{x^{\frac{2}{3}}}$$
$$= x^{-\frac{2}{3}}$$

$$y = \frac{x^{\frac{1}{3}}}{\frac{1}{3}} + c$$
$$= 3\sqrt[3]{x} + c$$

Question 10

$$\int (y+1)^4\,dy = \frac{(y+1)^5}{5} + c$$

Question 11

$$\int_0^1 (3x - 10)^{-2}\,dx$$
$$= \left[\frac{(3x-10)^{-1}}{3\times(-1)}\right]_0^1$$
$$= \left[\frac{-1}{3(3x-10)}\right]_0^1$$
$$= \left(\frac{-1}{3\times(-7)}\right) - \left(\frac{-1}{3\times(-10)}\right)$$
$$= \frac{1}{21} - \frac{1}{30}$$
$$= \frac{1}{70}$$

Question 12

$$\int \frac{dx}{(1\text{-}2x)^3} = \int (1-2x)^{-3}\,dx$$
$$= \frac{(1-2x)^{-2}}{-2\times(-2)} + c$$
$$= \frac{1}{4(1-2x)^2} + c$$

Question 13

$$\frac{x-1}{1-x}=\frac{x-1}{-(x-1)}$$
$$=-1$$

$$\int\frac{x-1}{1-x}\,dx=\int -1\,dx$$
$$=-x+c$$

Question 14

$$\int\frac{4x^2-4x+1}{4x-2}\,dx=\int\frac{(2x-1)^2}{2(2x-1)}\,dx$$
$$=\frac{1}{2}\int 2x-1\,dx$$
$$=\frac{1}{2}\left(\frac{2x^2}{2}-x\right)+c$$
$$=\frac{1}{2}(x^2-x)+c$$

Question 15

$$\int_{-1}^{1}\frac{x^5+x^3-x}{x^3}\,dx=\int_{-1}^{1}\frac{x^5}{x^3}+\frac{x^3}{x^3}-\frac{x}{x^3}\,dx$$
$$=\int_{-1}^{1}x^2+1-x^{-2}\,dx$$
$$=\left[\frac{x^3}{3}+x-\frac{x^{-1}}{-1}\right]_{-1}^{1}$$
$$=\left[\frac{x^3}{3}+x+\frac{1}{x}\right]_{-1}^{1}$$
$$=\left(\frac{1}{3}+1+1\right)-\left(\frac{-1}{3}-1-1\right)$$
$$=2\frac{1}{3}-\left(-2\frac{1}{3}\right)$$
$$=4\frac{2}{3}$$

Question 16

$$f'(x)=9\sqrt{x}+2$$
$$=9x^{\frac{1}{2}}+2$$

$$f(x)=\frac{9x^{\frac{3}{2}}}{\frac{3}{2}}+2x+c$$
$$=6\sqrt{x^3}+2x+c$$

$$f(2)=0$$

$$0=6\times\sqrt{8}+4+c$$
$$c=-6\times2\sqrt{2}-4$$
$$=-12\sqrt{2}-4$$

$$f(x) = 6\sqrt{x^3} + 2x - 12\sqrt{2} - 4$$

$$f(1) = 6 + 2 - 12\sqrt{2} - 4$$
$$= 4 - 12\sqrt{2}$$

Question 17

$$\frac{dy}{dx} = (x-2)(x+2)$$
$$\frac{dy}{dx} = x^2 - 4$$
$$y = \frac{x^3}{3} - 4x + c$$

$$y = 2,\ x = 2$$
$$2 = \frac{8}{3} - 8 + c$$
$$c = 7\frac{1}{3}$$

$$y = \frac{x^3}{3} - 4x + 7\frac{1}{3}$$

Question 18

$$\frac{d^2y}{dx^2} = -1$$
$$\frac{dy}{dx} = -x + c$$

$$\frac{dy}{dx} = x = -1$$
$$-1 = 1 + c$$
$$c = -2$$

$$\frac{dy}{dx} = -x - 2$$

$$y = -\frac{x^2}{2} - 2x + c$$

$$y = x = -1$$
$$-1 = -\frac{1}{2} + 2 + c$$
$$c = -2\frac{1}{2}$$

$$y = -\frac{x^2}{2} - 2x - 2\frac{1}{2}$$

Question 19

$$\int \sqrt{1+4x}\, dx = \int (1+4x)^{\frac{1}{2}}\, dx$$
$$= \frac{(1+4x)^{\frac{3}{2}}}{4\times\frac{3}{2}} + c$$
$$= \frac{\sqrt{(1+4x)^3}}{6} + c$$

Question 20

$$LHS = \left[\frac{x^5}{5} + x^3\right]_{-2}^{2}$$

$$= \left(\frac{32}{5} + 8\right) - \left(\frac{-32}{5} - 8\right)$$

$$= 28\frac{4}{5}$$

$$RHS = 2\left[\frac{x^5}{5} + x^3\right]_{0}^{2}$$

$$= 2\left[\left(\frac{32}{5} + 8\right) - 0\right]$$

$$= 28\frac{4}{5}$$

$$= LHS$$

BASICS OF INTEGRATION (CHALLENGE)

Question 1

$$\int (x^2+3)^2\,dx = \int (x^4+6x^2+9)\,dx$$
$$= \frac{x^5}{5}+\frac{6x^3}{3}+9x+c$$
$$= \frac{x^5}{5}+2x^3+9x+c$$

Question 2

$$g'(x) = 2x(x^2+3)^3$$

$$g(x) = \int 2x(x^2+3)^3\,dx$$
$$= \frac{2x(x^2+3)^4}{4\times 2x}+c$$
$$= \frac{(x^2+3)^4}{4}+c$$

Question 3

$$\int_1^4 x\sqrt{x}\,dx = \int_1^4 x^{\frac{3}{2}}\,dx$$
$$= \left[\frac{x^{\frac{5}{2}}}{\frac{5}{2}}\right]_1^4$$
$$= \left[\frac{2\sqrt{x^5}}{5}\right]_1^4$$
$$= \left[12\frac{4}{5}-\frac{2}{5}\right]$$
$$= 12\frac{2}{5}$$

Question 4

$$y = \sqrt{x^2-1}$$
$$= (x^2-1)^{\frac{1}{2}}$$

$$\frac{dy}{dx} = \frac{1}{2}(x^2-1)^{-\frac{1}{2}}\times 2x$$
$$= \frac{x}{\sqrt{x^2-1}}$$

$$\int_2^3 \frac{x}{\sqrt{x^2-1}}\,dx = \left[\sqrt{x^2-1}\right]_2^3$$
$$= \left[\sqrt{3^2-1}-\sqrt{2^2-1}\right]$$
$$= \sqrt{8}-\sqrt{3}$$
$$= 2\sqrt{2}-\sqrt{3}$$

Question 5

$$\pi\int_1^3 (x^4 + 4x^{-2})\, dx$$

$$= \pi\left[\frac{x^5}{5} + \frac{4x^{-1}}{-1}\right]_1^3$$

$$= \pi\left[\frac{x^5}{5} - \frac{4}{x}\right]_1^3$$

$$= \pi\left[\left(\frac{243}{5} - \frac{4}{3}\right) - \left(\frac{1}{5} - 4\right)\right]$$

$$= \pi \times 51\frac{1}{15}$$

$$= \frac{766\pi}{15}$$

Question 6

$$y = \sqrt[3]{x}$$

$$x = y^3$$

$$\int_1^8 x\, dy = \int_1^8 y^3\, dy$$

$$= \left[\frac{y^4}{4}\right]_1^8$$

$$= 1024 - \frac{1}{4}$$

$$= 1023\frac{3}{4}$$

Question 7

$$\frac{d}{dx}\left(\frac{x+1}{x-1}\right) = \frac{1(x-1) - 1(x+1)}{(x-1)^2}$$

$$= \frac{x-1-x-1}{(x-1)^2}$$

$$= \frac{-2}{(x-1)^2}$$

$$\int \frac{dx}{(x-1)^2}$$

$$= -\frac{1}{2}\int \frac{-2}{(x-1)^2}\, dx$$

$$= -\frac{1}{2}\left(\frac{x+1}{x-1}\right) + c$$

Question 8

$$\left|\int_{-1}^0 x(x+1)(x-2)\, dx\right|$$

$$= \left|\int_{-1}^0 x(x^2 - x - 2)\, dx\right|$$

$$= \left|\int_{-1}^0 x^3 - x^2 - 2x\, dx\right|$$

$$= \left|\left[\frac{x^4}{4} - \frac{x^3}{3} - x^2\right]_{-1}^0\right|$$

$$= \left|0 - \left(\frac{1}{4} + \frac{1}{3} - 1\right)\right|$$

$$= \left|\frac{5}{12}\right|$$

$$= \frac{5}{12}$$

Question 9

$$\frac{ds}{dt} = \frac{(2t^2 - 7t - 15)(2t + 3)}{t - 5}$$

$$= \frac{(2t^2 - 10t + 3t - 15)(2t + 3)}{t - 5}$$

$$= \frac{(2t(t - 5) + 3(t - 5))(2t + 3)}{t - 5}$$

$$= \frac{(t - 5)(2t + 3)(2t + 3)}{t - 5}$$

$$= (2t + 3)^2$$

$$s = \frac{(2t + 3)^3}{3 \times 2} + c$$

$$= \frac{(2t + 3)^3}{6} + c$$

Question 10

$LHS =$

$$2\int_{-1}^{0}(x^2 + 3x - 1)\,dx$$

$$+ 2\int(x^2 + 3x - 1)\,dx$$

$$= 2\left[\frac{x^3}{3} + \frac{3x^2}{2} - x\right]_{-1}^{0}$$

$$+ 2\left[\frac{x^3}{3} + \frac{3x^2}{2} - x\right]_{0}^{2}$$

$$= 2\left[0 - \left(-\frac{1}{3} + \frac{3}{2} + 1\right)\right]$$

$$+ 2\left[\left(\frac{8}{3} + 6 - 2\right) - 0\right]$$

$$= 2 \times (-2\frac{1}{6}) + 2 \times 6\frac{2}{3}$$

$$= 9$$

$$LHS = \int_{-1}^{2} 2(x^2 + 3x - 1)\,dx$$

$$= 2\left[\frac{x^3}{3} + \frac{3x^2}{2} - x\right]_{-1}^{2}$$

$$= 2\left[\left(\frac{8}{3} + 6 - 2\right) - \left(\frac{-1}{3} + \frac{3}{2} + 1\right)\right]$$

$$= 2 \times \left[6\frac{2}{3} - 2\frac{1}{6}\right]$$

$$= 9$$

WORKING

WORKING

WORKING

WORKING

WORKING

WORKING